HOW THE CLIMATE CRISIS CAN BE SOLVED

***

Instead of the failed ITER fusion project and its TOKAMAK principle

Åke Jean Hedberg

Publisher: BoD • Books on Demand, Stockholm, Sweden
Print: Libri Plureos GmbH, Hamburg, Germany
ISBN: 978-91-8057-688-8

Introduction

Let's begin by considering the common view of fusion energy and its application:

Fusion energy (commonly known as hydrogen power) is energy released by the fusion of light atoms. Energy production in the Sun and other main sequence stars is based on fusion. Fusion power plants are a hypothetical future for m of nuclear power plants, which would use fusion energy. (Research into fusion reactors began in the 1940s, but as of 2024, no device has reached net power, although net positive reactions have been achieved).

The advantage of fusion power plants over traditional nuclear power plants would be that the process does not have to leave behind as strongly radioactive substances as with fission. The problem with fusion is that extremely high temperatures must be able to be controlled, which is not possible with today's technology.

Instead of splitting heavy nuclei (fission), energy can be released through the fusion (combination) of light atomic nuclei using processes akin to the energy production in the Sun and other main sequence stars. No such power plants are yet in commercial operation, but research and development work is ongoing because the potential benefits are very large. Most people have been interested in the following reaction:

$$D + T \rightarrow 4He + n + 17.6\ MeV.$$

Most of the released energy consists of kinetic energy of the neutron that is released.

One way to achieve this fusion of deuterium and tritium is to heat the atoms to extremely high temperature (over 100 million degrees) and high pressure (8 atm). Since no materials can withstand such temperatures, they try to enclose the heated plasma in a magnetic field inside a torus-shaped tank. It could be a tokamak or a stellarator. So far, this can only be done for a very short time. The neutrons are unaffected by the magnetic field and hit the walls of the tank, which is covered by a blanket (a blanket) that absorbs the energy and where the heat is removed with a suitable cooling medium, for example water vapor or a gas such as helium. Another method is inertial confinement fusion, exposing small capsules of deuterium and tritium to intense laser, X-ray or particle pulses, whereby fusion processes can start. Up until now, it has also required the supply of more energy to run the process than could be extracted from it. A commercial exploitation of fusion power is probably between 30 and 50 years into the future at best.

On December 13, 2022, the United States Energy Administration announced that scientists at the National Ignition Facility had for the first time succeeded in achieving a net gain of energy from a fusion reaction, but that the energy conversion still requires more energy than is produced in the reaction.

The risk of catastrophic accidents similar to, for example, the Chernobyl accident is non-existent because the amount of fuel in the reactor is very small compared to a conventional nuclear power plant. It is expected that no one outside a fusion facility may need to be exposed to radiation; radiation protection is only needed for those who work at the plant. The D-T reaction does not give rise to radioactive waste, but material in the reactor construction can become radioactive. With an appropriate choice of construction material, the radioactive waste is relatively short-lived (up to approximately 100 years).

Tritium can be produced in the reactor from lithium-6 and lithium-7, whereby energy is also produced. Deuterium is found in seawater in large quantities and together with the lithium produced during the reaction there is a practically unlimited supply of material.[6] At the same time, some experts strongly question the realism of producing tritium in this way.

The International Thermonuclear Experimental Reactor (Iter) is an experimental facility for fusion research planned by several countries (EU, Japan, Russia, China, USA, South Korea, India). It is estimated to be completed in Cadarache in France in 2025, then used for approx. 20 years. The reactor will be cylindrical, 24 meters high and 30 meters wide, a so-called tokamak (meaning "annular magnetic chamber" in Russian).

The goal since 1985 has been to build ITER through international cooperation. The original participants were the European Union, Japan, the United States and the Soviet Union. The US dropped out of the process in 1998, but returned again in 2003. Later, Russia took over the role of the Soviet Union, and South Korea, China and India joined as new partners.

The project underwent a protracted process to determine the facility's location. Cadarache was supported by the EU, Russia and China, while the US, Japan and South Korea generally thought Rokkasho in Japan was a better place. Both sites were suitable for construction, and only political opinions guided the opinions. Some have argued that US opposition to building in France is based on French criticism of the Iraq War.[3] The agreement to locate the facility in France was reached on June 28, 2005.

In 2006, work began on preparing the area where the facility is to be built. Construction of ITER itself began in 2009 and is expected to be completed in 2025. On July 28, 2020, a ceremony was held at ITER's assembly hall when the construction of the Tokamak itself began. to be completed in 2025.

The cost of building the facility was estimated in 2016 at 20 billion euros. [6] According to the plan in 2005, the facility was to be completed in 2016, but has been rescheduled and is expected to be completed in 2025.

The Joint European Torus, abbreviated JET, was a pan-European fusion research facility in Culham in Great Britain with the world's largest tokamak reactor to date. JET was the world's largest and most successful experimental facility for fusion research. In 1970 the Council of the European Union decided that it should be built, three years later construction started. In 1978 it was completed on time and on budget. The plant was in operation until December 2023.

In 1991, a big step was taken in fusion research when, on November 9, fusion was created for the first time in a controlled manner. In 1997, it created 16 megawatts of fusion energy after adding 24 megawatts. It then returned 65% of the energy used. The facility was operated as a European collaboration within the framework of the Euratom program which is part of the EU's framework program for research.

In the 2000s, JET was upgraded in several rounds, including with new materials in the vessel wall. On December 21, 2021, they managed to keep a hot plasma contained for five seconds, releasing 59 megajoules of energy. This was a new record in fusion research, although more energy had to be added than released. In the plant's last test in October 2023, even more was released, 69 megajoules.

The new ITER research facility is expected to be completed in 2025. It then becomes the largest and most efficient fusion reactor. The results from JET support the development of ITER.

We can notice three facts here:

1)Energy production in the Sun and other main sequence stars is based on fusion. Fusion power plants are a hypothetical future form of nuclear power plants, which would use fusion energy.

2) Research into fusion reactors began in the 1940s,

3) but as of 2024, no device has reached net power, although net positive reactions have been achieved.

I ask my AI assistant:

What takes it to make a fusion power plant work. What is important?

In order to make a fusion power plant work, it is necessary to be able to control and reproduce fusion reactions that take place in the sun. Fusion reactions occur when

To make fusion power plants work, it is important to be able to create and maintain high enough temperatures (several million degrees) and pressure (several billion atmospheres) for the nuclei to fuse together and release energy. It is also important to be able to maintain a controlled fusion of fuel, usually deuterium and tritium, and to be able to extract the energy efficiently.

Other important factors in making a fusion power plant work include being able to handle high neutron fluxes and radioactive waste generated by fusion, as well as having an efficient cooling and energy transfer system to keep the reactor and its components stable and operational. With continued research and development, it is hoped to make fusion power plants more practical and economically profitable in order to contribute to a sustainable energy supply in the future.

My AI assistant can only explain the common and universally recognized view - the theory - of how the fusion processes on the sun and the other stars work and thus how these should, can and must be imitated here on earth in a special fusion reactor facility. And as we can see, it is not particularly sophisticated: brutal violence is what counts: Extremely high temperature and ditto pressure:(several million degrees) and pressure (several billion atmospheres)….och tekniken måste bli därefter: and the technique must be accordingly.

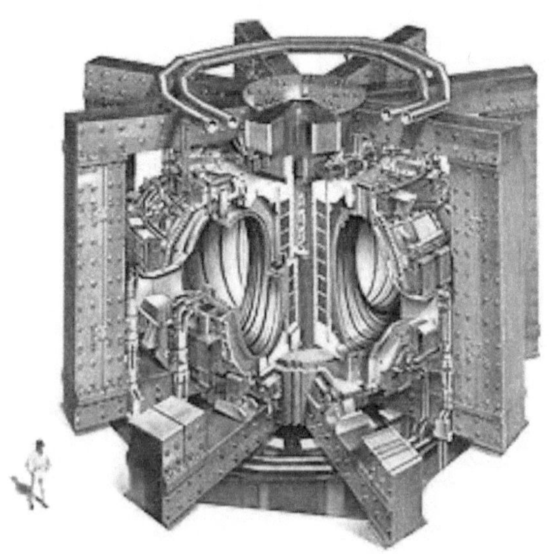

What happens on the sun they mean is this:

The energy from the Sun - both heat and light energy - originates from a **nuclear fusion process** that is occurring inside the core of the Sun. The specific type of fusion that occurs inside of the Sun is known as **proton-proton fusion.**

Inside the Sun, this process begins with protons (which is simply a lone hydrogen nucleus) and through a series of steps, these protons fuse together and are turned into helium. This fusion process occurs inside the core of the Sun, and the transformation results in a release of energy that keeps the sun hot. The resulting energy is radiated out from the core of the Sun and moves across the solar system.[3] It is important to note that the core is the only part of the Sun that produces any significant amount of heat through fusion (it contributes about 99%). The rest of the Sun is heated by energy transferred outward from the core.

The overall process of proton-proton fusion within the Sun can be broken down into several simple steps. A visual representation of this process is shown in Figure 1 (above). The steps are:

1. Two protons within the Sun fuse. Most of the time the pair breaks apart again, but sometimes one of the protons transforms into a neutron via the weak nuclear force. Along with the transformation into a neutron, a positron and neutrino are formed. This resulting proton-neutron pair that forms sometimes is known as deuterium.
2. A third proton collides with the formed deuterium. This collision results in the formation of a helium-3 nucleus and a gamma ray. These gamma rays work their way out from the core of the Sun and are released as sunlight.
3. Two helium-3 nuclei collide, creating a helium-4 nucleus plus two extra protons that escape as two hydrogen. Technically, a beryllium-6 nuclei forms first but is unstable and thus disintegrates into the helium-4 nucleus.

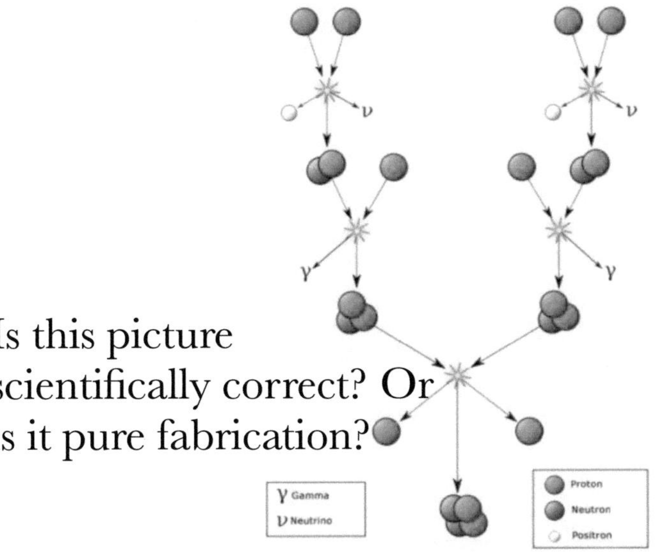

Solar neutrino problem, long-standing astrophysics problem in which the amount of observed neutrinos originating from the Sun was much less than expected.
In the Sun, the process of energy generation results from the enormous pressure and density at its centre, which makes it possible for nuclei to overcome electrostatic repulsion. (Nuclei are positive and thus repel each other.) Once in some billions of years, a given proton (1H, in which the superscript represents the mass of the isotope) is close enough to another to undergo a process called inverse beta-decay, in which one proton becomes a neutron and combines with the second to form a deuteron (2D). This is shown symbolically on the first line of equation (1), in which e− is an electron and ν is a subatomic particle known as a neutrino

This is yet another hasty result from the theory that enormous pressure and density at its center, which one only BELIEVE is required for fusion! In my theory the Sun need neutrinos, like we need oxygen.

The final helium-4 atom has less mass than the original 4 protons that came together (see E=mc2). Because of this, their combination results in an excess of energy being released in the form of heat and light that exits the Sun, given by the mass-energy equivalence. To exit the Sun, this energy must travel through many layers to the photosphere before it can actually emerge into space as sunlight. Since this proton-proton chain happens frequently - 9.2×10^{37} times per second - there is a significant release of energy.[3] Of all of the mass that undergoes this fusion process, only about 0.7% of it is turned into energy. Although this seems like a small amount of mass, this is equal to 4.26 million metric tonnes of matter being converted to energy per second. Using the mass-energy equivalence, we find that this 4.26 million metric tonnes of matter is equal to about 3.8×10^{26} joules of energy released per second!

***

About fusion processes on the sun and what you should know.

Nature and society are our main sources of knowledge. In nature we find, therefore, all the processes and mechanisms we can use for our needs. This applies especially to the processes on the Sun if we want to "tame" its fusion energy.

"Generally, the proton proton fusion take place only if the temperature (the kinetic energy) of the protons is high enough to overcome their mutual forces created by Coulomb's law. The theory that proton-proton reactions were the basic principle behind the sun and other stars combustion was developed by Arthur Stanley Eddington in the 1920s. "(Wikipedia).

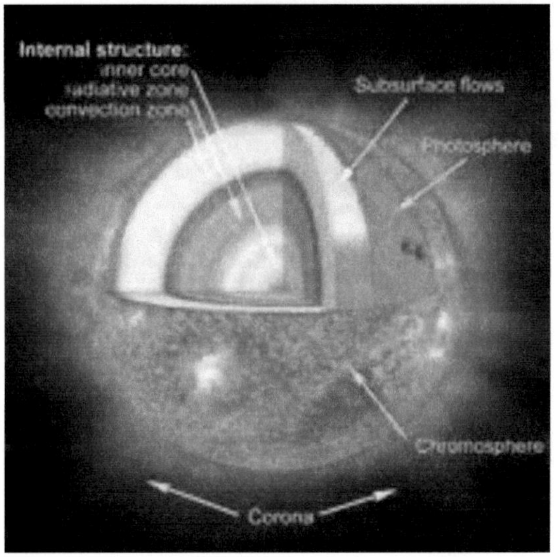

The innermost layer of the sun is the core. With a density of 160 g/cm^3, 10 times that of lead, the core might be expected to be solid. However, the core's temperature of 15 million kelvins (27 million degrees Fahrenheit) keeps it in a gaseous state.

In the core, fusion reactions produce energy in the form of gamma rays and neutrinos. Gamma rays are photons with high energy and high frequency. The gamma rays are absorbed and re-emitted by many atoms on their journey from the envelope to the outside of the sun. When the gamma rays leave atoms, their average energy is reduced. However, the first law of thermodynamics (which states that energy can neither be created nor be destroyed) plays a role and the *number* of photons increases. Each high-energy gamma ray that leaves the solar envelope will eventually become a thousand low-energy photons.

The neutrinos are extremely nonreactive. To stop a typical neutrino, one would have to send it through a light-year of lead! Several experiments are being performed to measure the neutrino output from the sun. Chemicals containing elements with which neutrinos react are put in large pools in mines, and the neutrinos' passage through the pools can be measured by the rare changes they cause in the nuclei in the pools.[3]

Those who still after almost 90 years of research claiming that the theory of "proton-proton reactions (is) the basic principle behind the sun and other stars burning" have not thought about and studied the matter thoroughly, I guess. I mean, that the sun's core activity is not a directly burning of protons into helium and heavier elements, like the conventional doctrine says. At first, it is instead the production of neutrons (and energy). The sun's method to solve the problem of the Coulomb barrier is then crucial to any attempt to construct a fusion machine. To overcome the Coulomb barrier is not happening as you maybe think through violence but through, if I may say, a very cleverer process, as you will see.[4] The current erroneous theory and the perception of this is the main cause of all previous failures to solve the energy issue. Do you really want to know the solution to the energy question and the principle that, so to speak, tame the fusion energy as the sun and the other stars develops? Then you should consider the following things:

---

[3] http://www.universetoday.com/40631/parts-of-the-sun/

[4] So particles with 3-10 keV of energy (which there are plenty of in the Sun's core) can overcome the Coulomb barrier. Voila! Nuclear Reactions!
http://burro.astr.cwru.edu/Academics/Astr221/StarPhys/coulomb.html

a. First, forget everything you perhaps know or have learned about quarks. Forget therefore u-quarks, d-quarks, strange-quarks, anti-quarks, etc. inventions that not logically could be derived from other phenomena.

b. Do not turn back to the tank and the old picture of how an electron, a proton or neutron are structured, ie as small incomprehensible dot-shaped beads. Equally incomprehensible and structure solved shaped as quarks, an "invention" that does not add anything new to the understanding. Only added new problems as "glue particles" (gluons), etc. Realize that you probably do not have a working image how these particles are structured. There is no understandable theory which goes to visualize.

c. Instead, take this to you: It's all about the well-known and well researched photons and neutrinos and their anti-forms. The electron is a combination of a neutrino and a photon, a proton a combination of two neutrinos and a antineutrino, a neutron of a proton plus a photon and its antiform. All this gives a whole new into every detail comprehensible structure, a new theory and model of the atom, of its core, shell and functioning. When this is done, you may go to:

I. Understanding neutron decay.
II. Solving the riddle of the fusion processes on the sun and why it still are on.
III. Learn the art of building a workable fusion machine by applying and controlling these forces of nature here on earth and thus solve the energy issue.

With this questioning of the basic theories and approaches, we are not alone. We have world-class naturalist on our side, such as Albert Einstein, Karl Popper and Hannes Alfvén. Some quotes:

All these fifty years of conscious brooding have brought me no nearer to the answer to the question, 'What are light quanta?' Nowadays every Tom, Dick and Harry thinks he knows it, but he is mistaken.[5]

Einstein /.../, believed that there must ask a further, deeper level in physics, a level beyond quantum mechanics.[6]

Since thermonuclear research started with Zeta, Tokamaks, Stellators – not to forget the Perhapsatron – plasma theories have absorbed a large part of the energies of the best physicists of our time. The progress which has been achieved is much less than was originally expected. The reason may be that from the point of view of the traditional theoretical physicist, a plasma looks immensely complicated. We may express this by saying that when, by an immense number of vectors and tensors and integral equations, theoreticians have prescribed what a plasma **mus**t do, the plasma – like a naughty child – refuse to obey. The reason is either that the plasma is so silly that it does not understand the sophisticated mathematics, or it is that the plasma is so clever that it finds other ways of behaving, ways which the theoreticians were not clever enough to anticipate. Perhaps the noise generation is one of the nasty tricks the plasma uses in its IQ competition with the theoretical physicists. /.../What is urgently needed is not a refined mathematical treatment /.../ but a rough analysis of the basic phenomena.[7]

---

5 Albert Einstein, in a letter to his old friend MA Besso, 1954.

6 Karl Popper, *Quantum Theory And The Schism in Physics.*

7 Hannes Alfvén. Opening lecture at the *"Double Layers and circuits in astrophysics"*, Marshall Space Flight Center, Huntsville, Alabama, March 17-19, 1986. TRITA-EPP -86-04. Department of Plasma Physics, The Royal Institute of Technology, Stockholm Sweden.

A workable fusion power? Yes, thanks!

Fusion research goes back to the 20s and 30s when researches tried to tame the energy that gets sun to shine. With their new physic and quantum mechanics many thought they completely understood the processes that occur on and in the sun. In the early 1950s they start their projects. And are still going on. And no functional fusion power plants is in sight the next few decades, the truth is that scientists do not know how to do to create the necessary conditions for precious ITER project.[8] They do not want, cannot and dare not face the truth that it is the theory that is wrong. As Hannes Alfvén - Nobel Laureate in plasma physics and an expert in both fusion technology and theory - pointed out already in the 1980s.

ITER, then, is to "tame solar energy" (or the hydrogen bomb) through a controlled fusion (not sudden and explosive), among other things, the hydrogen isotopes deuterium, tritium or isotopes of helium. Simply stated, it is a question of the opposite of fission, the splitting of certain heavy atoms such as uranium, thorium, plutonium, etc., thus releasing energy. Here it is instead merging of light isotopes of hydrogen, helium, lithium, etc.

The project itself is very commendable: to try to tame the solar energy. It is very much needed and time is running out. For in the fusion process it will not be any radioactive waste to a greater extent. No risk of meltdowns or the like and the raw material is in principle only plain tap water (for the fusion of helium isotopes are much to download on the Moon).The problem however is that they will apply in principle the same old hopeless tokamak technology already tested in nearly sixty – 60 – years! A technology that despite prolonged and persistent experiments have failed. It simply does not work. And why not? Well, as we may now understand it is a lack of understanding of what actually happens on the sun, and in general about the nature of reality. They are walking on in their old wool stockings! There are absolutely no new thoughts, new ideas or theories here.

But awareness of this lack of functional theories have still been around since long. For example, in one of Europe's top plasma physicist Hannes Alfvén in his time already decades ago warned of this deficiency. He has written a lot of this. He was in experiments similar tokamak technology already shortly after the war in the former Soviet Union. And the plasma he and other researchers using the same technology or variants of it since worked on, inter alia, KTH in Stockholm was too chaotic and erratic. It behaved (as we have seen) "like a disobedient child — it refused to obey". He concluded that what was needed was a "rough analysis of the fundamental phenomena of /.../ not a refined mathematical treatment". As Professor and Nobel Laureate in Physics in 1970, he knew.

---

8 The ITER Agreement was signed by **China**, the **European Union**, **India**, **Japan**, **Korea**, **Russia** and the **United States**. The Members of the ITER Organization will bear the cost of the project through its 10-year construction phase and its 20-year operational phase before decommissioning. In Cadarache, southern France, ITER construction has begun on the scientific facilities. Manufacturing of components for ITER is underway in Members' industries all over the world; the level of coordination required for the successful fabrication of over one million parts for the ITER **Tokamak** alone is daily creating a new model of international scientific collaboration. *https://www.iter.org/proj#collaboration*

Therefore opposed Alfvén the conventional nuclear power – fissionskraften – so strong (became the swedish prime minister Fälldin's scientific adviser in the nuclear debate on in the 1980s). One should make an effort to solve the basic theoretical and philosophical questions first, he said. The conventional nuclear power as a kind of *stopgap measure* while waiting for a proper functioning fusion technology, he could not accept.

Now he is long dead, but his insight about the lack of a sensible theory for these fusion processes is even more true today. Since the problem is demonstrably still and scandalous unresolved. But still it is "urgently needed" with a "rough analysis of the basic phenomena". The physicists can demonstrably after about sixty years do not claim that what is needed is "further research and experimentation". Everything is literally already tried which Alfvén could testify as early as the 1980s. What is therefore needed is a completely new theory, a new vision of our nature and reality, and thus a brand new technology built on entirely new principles than the construction of the "Perhapsatron" or the ITER which now being built in France. Tragically prevent this expensive hopeless building practical solution to the energy issue. While the resources and money to no avail runs out. And "the best physicists of our time" wasting their time.

*

So I am not the only or the first who points out the lack of theory. It has taken me many years, but now I think I know how to do it. Hope this compendium describe how it all can go to in order to bring about a controlled and from an energy profitable process. But first, a description from KTH in Stockholm in Sweden and Wikipedia what it is about:

> Fusion energy is the energy released by the fusion of light atoms. Energy production in the Sun and other main sequence stars based on fusion. Fusion Reactor is a hypothetical future form of nuclear power plants that would use fusion energy.
>
> The advantage of fusion power plants over traditional nuclear plants would be that the process does not need to leave behind as strongly radioactive substances fission. The problem with fusion is that extremely high temperatures must be monitored, which did not succeed in today's technology. Instead of splitting heavy nuclei (fission), the energy released by the fusion (merging) of light nuclei with processes that are related to energy production in the sun and other main sequence stars. No such power plants are yet in commercial operation but there is ongoing research and development as the potential benefits are enormous. Most have been interested in the following reaction:

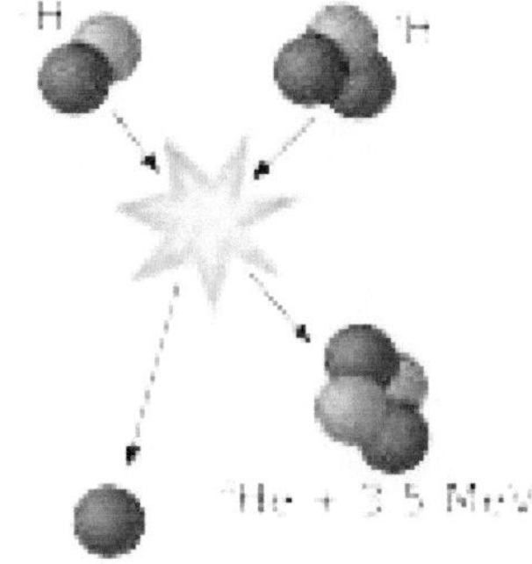

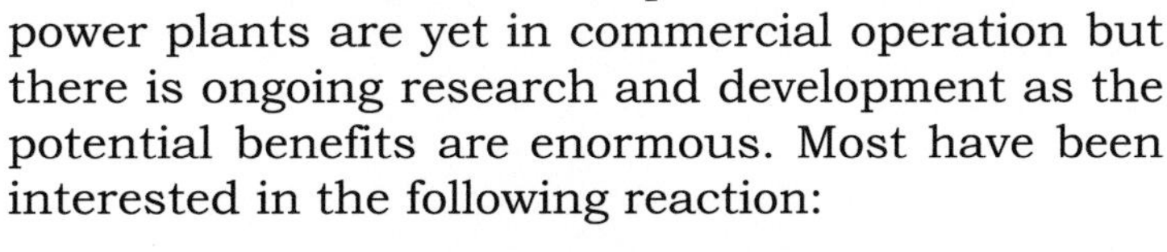

$$D + T \rightarrow {}^4He + n + 5.2 \times 10^{-13} J$$

or

$${}^2H + {}^3H \rightarrow {}^4He + 3{,}5\ MeV + n + 14{,}1 MeV$$

Most of the energy consists of kinetic energy of a neutron is liberated. One way to achieve this fusion of deuterium (D) and tritium (T) is to heat the atoms to extremely high temperature (over 100 million degrees) and high pressure (8 atm). Since no material can withstand such temperatures are trying to turn inside the heated plasma in a magnetic field inside a toroidal tank, called a tokamak. So long can handle it just this for a very short time. The neutrons are unaffected by the magnetic field and hits the tank walls that are covered of a blanket that takes up energy and where heat is removed with suitable cooling medium, such as water vapor or a gas such as helium. Another method is to bombard a hydrogen prepared with high-power lasers from all sides to the extreme compression, whereby it by adding a further laser pulse can ignite the process.

So far, there have also been required supplying more energy to run the process than might have been extracting from it. The commercial exploitation of fusion power is, at best, probably between 30 and 50 years into the future. The risk of catastrophic accidents like Chernobyl accident, for example, is nonexistent because the amount of fuel in the reactor is very small compared with a conventional nuclear power plants. It is estimated that no one outside of a fusion plant may need to be exposed to radiation without radiation protection is only necessary for those who work at the plant. DT reaction does not produce radioactive waste but the materials in reactor design can become radioactive. With appropriate choice of structural materials becomes the radioactive waste relatively short-lived (up to approximately 100 years).

How a conventional fusion power plant of the tokamak type:
(See the bibliography below)
Tritium can be produced in the reactor from lithium-6 and lithium-7 which also energy is produced. Deuterium is in sea water in large quantities and with available lithium has been estimated that fusion based on these isotopes would be enough for humanity in practically unlimited time (one million years [clarify]). At the same time questioning some experts strongly that realism to produce tritium in this way.

Containment
Although the above temperatures are high, the following can be found. Researchers have in exceptional circumstances reached up to temperatures around 500 million degrees Kelvin, which is five times more than what is needed in a fusion power plant. Part of the solution to the problem of enclosing the fuel lies in the fact that at the above high temperatures separated nuclei and electrons from each other. This is called ionization and the positively charged nuclei behave like ions. The hot gas which contains free negatively charged electrons and positively charged ions called plasma. Because of the electric charges which are electrons and ions in plasma may be contained in a magnetic field. In the absence of a magnetic field moves charged particles in the plasma in straight lines and irregular directions. Since nothing prevents the charged particles' movements, these can meet the enclosing walls which makes the plasma cools down and fusion reactions are prevented. But in a magnetic field that forced the charged particles follow the magnetic field lines of force. That is to say; the charged particles in the hot

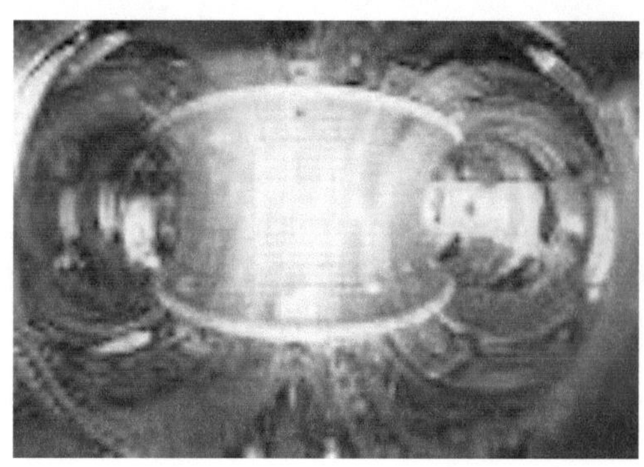

Interior of a Tokamak

plasma is contained in a magnetic field and thus prevented from hitting the enclosing walls.

Heating of the plasma.

In a working fusion reactor, part of the energy released will be used to maintain the plasma temperature, as new deuterium and tritium entering. However, the plasma must be heated to 100 million degrees Kelvin at startup, or after a temporary halt. In the current magnetic fusion experiments are not obtained enough energy to maintain the plasma temperature. Consequently, as they operate in short pulses and plasma must be heated again before each pulse.

Resistive heating plasma.

Resistive heating plasma may be heated by a current conducted through it. This is made possible by the plasma is an electrical conductor. This heating is called resistive heating and is the same as happens in a light bulb when a current is passed through it. This can be done in two ways. Either by inducing a current in the plasma (by changing the magnetic field) or with microwaves. The latter is done by plasma irradiated with powerful microwaves transport of electrons. The temperature achieved depends on how much resistance is in the plasma and how strong the current is. As the temperature is increased in the plasma the plasma resistance decreases. This allows the maximum temperature that is reachable using resistive heating, is 20-30 million degrees Kelvin. To achieve higher temperatures, other heating methods used.

Magnetic compression

A gas can be heated by compression. Similarly, the plasma is compressed and the temperature in the raised by rapidly increasing the surrounding magnetic field. In a tokamak is done by moving the plasma to a region of higher magnetic field (reduced radius) A side upon compression of the plasma is that the ions are closer together, with the result that the density increases. High density is one of the basic requirements for fusion to take place.

Neutral Beam Injection

Deuterium and tritium is pushed into the plasma high velocity. Once inside the plasma ring ionized atoms, after which the ions are slowed down and the kinetic energy transferred to the plasma.

Radio waves

When heating of the plasma by means of radio waves so generated a wave of oscillators outside the chamber. The waves have a certain frequency or wavelength is selected to match the ions spiraling around magnetic field lines in the plasma ring and wave energy is transferred to the charged particles in the plasma. These in turn collide with other particles, causing the temperature of the plasma increases.

A likely method in a fusion power plant of the tokamak type is that a deuterium-tritium mixture is introduced into the reactor chamber. The fuel is separated from the chamber walls by means of a very strong magnetic field. The magnet coils generate a "torodialt" magnetic field. With the transformer induced a strong current in the plasma. The current creates a "polodialt" magnetic field inside the machine. Together, the two magnetic fields a helical magnetic field that keeps the hot plasma in place.

When the plasma reached a high enough temperature, the merger process started. It's called reactor "teeth". Fusion energy is released and the kinetic energy of the neutrons captured in a cooling medium (water, helium gas or liquid sodium) contained in a jacket in the reactor wall. The coolant for the energy out of the reactor through heat exchangers and turbines, in the same way as in today's nuclear power plants. The sheath also has a sheath of lithium irradiated with neutrons thus produced reactor own tritium fuel. What keeps the fusion process at the lighthouse in a future fusion reactor, the energy of the so-called alpha particles (helium ions). As soon as they surrendered their energy to the plasma must be removed in order not to dilute the fuel and disrupt the fusion process. This can be done through a so-called "divertor" that controls the magnetic field on the plasma ring surface. Helium ions is then discharged from the reactor chamber via a pipe in the ground.

Fusion environmental

The first fusion reactors will not be completely free from radioactivity. The strong neutron stream in the reactor core contaminates the walls which means it requires long-term storage of this material. In future reactors can however use the material which has a different structure which reduces the radioactivity at levels far below today's nuclear reactors. Even a fusion reactor will produce nuclear waste as well as today's nuclear power plants. However, there is some difference in this radioactive waste. Today's nuclear waste decays in ten thousand years, while the nuclear waste from a fusion reactor decay of 100 years. Another materiality is all radioactive fuel can be created in the reactor of non-radioactive material. Transport of radioactive uranium is not needed. In addition, a fusion reactor does not suffer a meltdown, but turns itself off if problems occur.

Future plans

It's now functioning fusion machine is experimental reactor JET (Joint European Torus) in England. However, it is brought out and will be closed in December 1999. Nextstep in the research is to build ITER (International Thermonuclear Experimental Reactor). It was scheduled to begin in 1999, but is indefinite cry when combined price tag of 60 billion discouraging. ITER is as JET, a "tokamak" but it is 20 times larger. ITER is meant to be a research facility on the road to a real fusion power plant. It should be "light" and "burn" in more than one thousand-seconds, and it shall test modules for manufacturing its own tritium fuel. The step after ITER is that in 2020 to build a demonstration reactor to produce electricity. Only then, in fifty years, perhaps the first commercial fusion power plant sees the light of day.

References http://www.e.kth.se/~e97_bli/fusion.html
Books Hagler, Marion O. An introduction to controlled thermonuclear fusion, 1977
Journals New Technology 1998: 17 Science Illustrated 1995: 2

More ITER:

ITER (Acronym of the International Thermonuclear Experimental Reactor) is an experimental facility for research on fusion projected by several countries (EU, Japan, Russia, China, the US, South Korea, India). It is scheduled for completion in Cadarache in France in 2018. [1] The reactor will be cylindrical, 24 meters high and 30 meters wide, called a tokamak (means in Russian "annular magnetic chamber").

The goal has since 1985 been to build ITER through international cooperation. The original participants were European Union, Japan, the United States and the Soviet Union. US jumped in 1998 by the process, but came back again in 2003. Later, Russia has taken over the role of the Soviet Union, and South Korea, China and India come up with new partners.

The project went through a lengthy process to determine the plant's location. Cadarache was supported by the EU, Russia and China, while the US, Japan and South Korea generally thought to Rokkasho in Japan was a better place. Both sites were suitable for construction, and only political views that ruled opinions. Some have suggested that US opposition to building in France is based on the French criticism of the Iraq war. [2] The agreement to place the plant in France reached 28 June 2005.
In 2006 began a user to work with to prepare the area facility will be built on. The construction of the ITER started in 2009 and expected to be completed until 2018.

Cost.

The cost to build the facility was estimated in 2005 to 4.7 billion euros. A new report says that the price could be up to 30% higher. The plant was planned in 2005 for completion in 2016.

Fuel

The fuel in the power plant is planned to be hydrogen forms deuterium and tritium, which collide with each other and form helium when the temperature reaches 10^8 ° C.

The energy is divided so that the neutron get 14 MeV and 3.6 MeV helium gets into kinetic energy. This is the same reaction used in nuclear weapons. But since the release of neutrons is very hard to deal with (a part used to produce new tritium in the reactor, the other sits on the reactor walls and form radioactive material), the helium-3 proposed. Hydrogen forms are available in unlimited quantity, but it is worse with helium-3.

References

^ The Energy Daily: Nuclear fusion power project to start in 2018: official (taken 20100224)
^ BBC News: Fusion Project Decision delayed (20 December 2003)MG Axelsson, "Fusion - closer than ever," Daily News (7 March 2004)
http://www.dn.se/nyheter/vetenskap/fusion-narmare-an-nagonsin-1.249759
M. Scott, D. Johnson, "Science Matters - Nuclear Power" (1997)
external Links
http://www.nature.com/news/2008/080612/full/453829a.html
Retrieved from
"http://sv.wikipedia.org/w/index.php?title=ITER&oldid=15680531"
Categories: Quality 2010-11Artiklar who need occasional sources 2010-10Kärnteknik

*

A first summary:

"The problem with fusion is that extremely high temperatures must be monitored, which did not succeed in today's technology." No, and will not reasonably have ever succeed. No, tokamak technology is doomed. The root cause is thus (quote from page 10):

The truth is that one does not know how to do to create the conditions necessary for a successful merger. They do not want, cannot and dare not face the truth that it is the theory that is wrong. It's not even part of the training of these nuclear physicists and fusion scientists, so they have no chance to even understand that. Yes, they don't even understand that they don't understand. They just believe it.... Which is really bad.

This has Hannes Alfvén - Nobel Prize in Plasma Physics and expert on fusion technology — pointed out already in the 1980s. An indirect recognition of these state of affairs is that *The commercial exploitation of fusion power is, at best, probably between 30 and 50 years into the future.*

In the best case 30 years but probably 50! What is needed then? Well, as Alfvén argued, required a new theory that allows a significantly lower operating temperature than the Top 100 million degrees now assumed to be the correct one. Or rather a correct theory of how the merger actually goes in reality and in nature. Moreover there is something wrong with the theory which is assumed to apply to the sun.

Evidently, we must leave the technology with the magnetic inclusions of type TOKAMAK behind us and the 100 millions of degrees.
So there is something wrong with the prevalent theory of the mechanisms that generate solar energy. The sun radiates, only a third as much solar neutrinos as it should according to the conventional theory. Solutions to the solar neutrino problem are usually classified into one of two categories, astrophysical and physical. Solutions that require a change in how we see the sun called astrophysical solutions, while solutions that require a change in how we think neutrinos are called physical solutions. My solution is that there is something wrong with the theory of what actually happens on and in the sun.

The basic idea of this (old) theory is that two hydrogen atoms (protons) directly fused into deuterium (see if the proton-proton chain below) whe- reby large amounts of energy are formed. This, in a first step. But this has created the so-called neutrino problem. My idea is that ***neutrons*** are formed in a first step.Which solved the problem in question.
. The neutrino problem usually classified into one of two categories, astrophysical and physical. Solutions that require a change in how we see the sun called astrophysical solutions, while solutions that require a change in how we think neutrinos are called physical solutions. My solution is that there is something wrong with the whole theory of what actually happens on the sun. Basically it's about our view of nature and how we can best imitate its processes.

Proton-proton chain reaction (From Wikipedia)

Proton-proton chain reaction is one of several fusion reactions by which stars convert hydrogen to helium, the main option is the CNO cycle. Proton-proton chain dominates in stars of the sun's size or smaller.
To overcome the electrostatic repulsion between two hydrogen nuclei requires a large amount of energy and this reaction takes an average of 109 years for it to be completed at the temperature of the sun's core. Because of the slow reaction is still shining sun; had it been rapid sun would have exhausted its hydrogen long ago.

Generally, the proton-proton fusion take place only if the temperature (the kinetic energy) of the protons is high enough to overcome their mutual forces created by Coulomb's law. The theory that proton-proton reactions were the basic principle behind the sun and other stars combustion was developed by Arthur Stanley Eddington in the 1920s. At the time, however, considered the temperature of the sun may be too low to overcome the coulomb barrier. The development of quantum mechanics, however, opened soon for the theory, when it was discovered that the tunneling effects could allow fusion at lower temperatures than those predicted by classical physics.

Proton-proton reaction

The first step involves the fusion of two hydrogen nuclei 1H (p + = proton) to the deuterium (D), releasing a positron (e^+) and a neutrino (v_e) because one proton changes into a neutron.

$$p^+ + p^+ \rightarrow 2D + e^+ + ve + 0.42 \text{ MeV}$$

This first step is extremely slow, both because the protons must tunnel through the Coulomb barrier and since it depends on the weak interaction. The positron annihilate immediately with an electron, and their mass energy is carried away by two gamma photons.

$$e- \quad + \quad e- + \rightarrow 2\gamma + 1.02 \text{ MeV}$$

After this, the deuterium created in the first step merge with the hydrogen nucleus and form a light isotope of helium, 3He.

$$2D + 1 H \rightarrow 3He + \gamma + 5.49 \text{ MeV}$$

From this point, there are three possible ways to form helium isotope 4He. But this we can ignore for the moment, to name just a reaction to:
Pep reaction Deuterium can also be formed by the unusual peptide reaction (proton-electron-proton).

$$p^+ + e^- + p^+ \rightarrow 2D + v$$

(where p^+ is a proton, e^- is a negative electron and v_e is a electron neutrino).

The sun takes p-p reaction about 400 times more often than pep reaction. But neutrinos created by the peptide reaction is more energy dense. While they created in the first step of the p-p reaction has an energy up to 0.42 MeV gives pep reaction is a sharp energy line at 1.44 MeV. PEP and p-p reactions can be seen as two different Feynman representations of the same basic reaction, in which the electron passes to the right side of the reaction as an anti-electron. (End of quote from Wikipedia.)

Before we move on to my idea, let's look closer at the beta-decay, specifically neutron decay in two stages; a process that still caused and are causing physicists much headache. The understanding of this process is essential in order to understand why the sun (still) is burning and shines. Which in turn is essential for finding the solution to the fusion problem here on Earth.

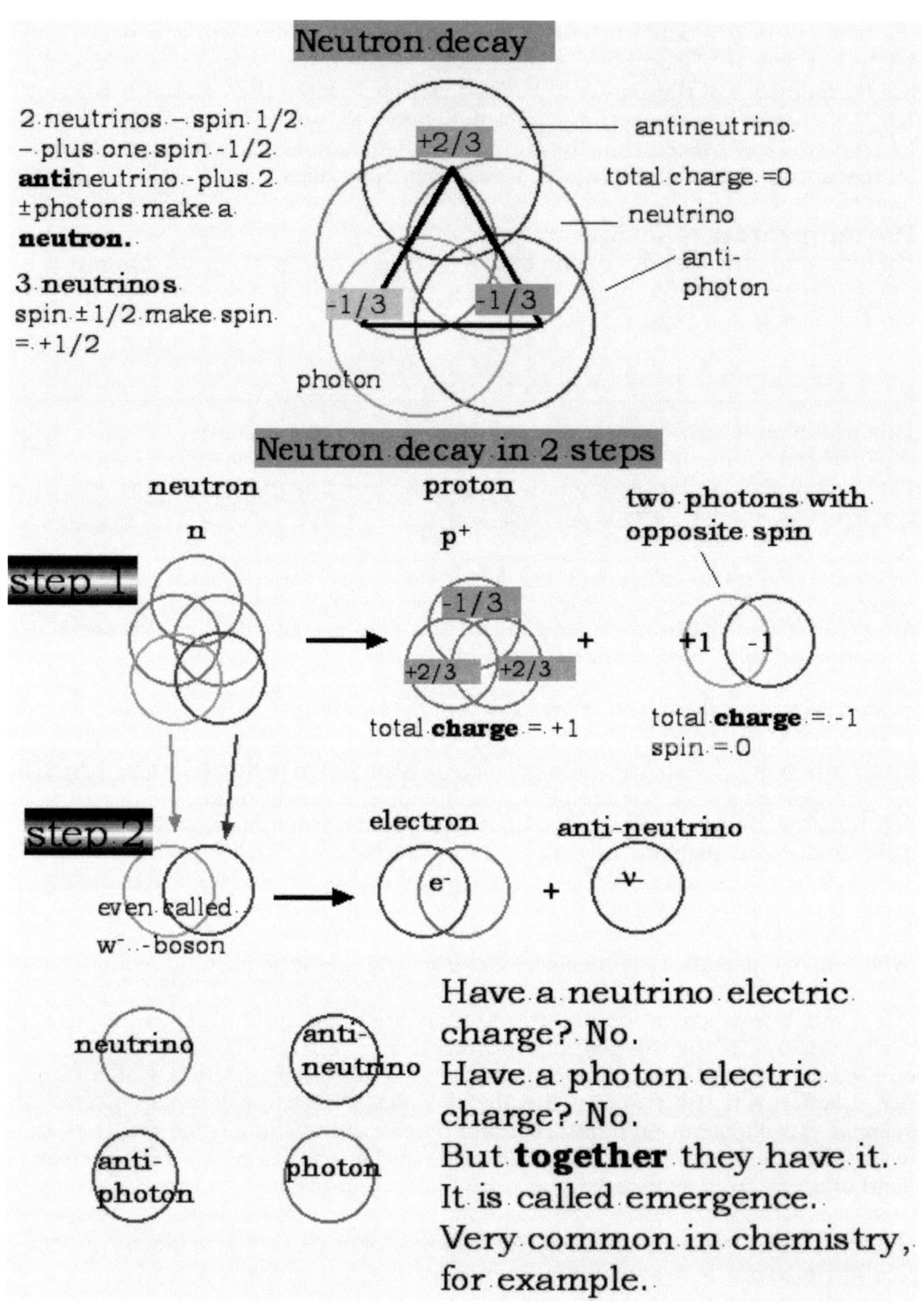

But this epistemological truth is not part of the usual academic, natural science education. Unfortunately. But this lack of knowledge is not understood. They don't understand that they don't understand!

The different colored circles symbolize thus photons (yellow) and neutrinos (red) plus their anti-forms (blue and green).

The quarks are here replaced by the well-known neutrinos (up-quark) and anti-neutrinos (down), one might say. This theory thus has the full support of the actual processes that here neutron decay, but also thousands and thousands of experiments and observations.

My basic idea about what is actually happening on the sun and why it burns and shines as it does is to say that neutrons are formed in a first step. Then merge the neutron with a proton. Then we get deuterium (D) which means that the rest is easy. Now is the way open for the sun to produce the energy we know and that makes it shine! So ***instead*** of the reactions described above and which we now term reactions (1) and (2):

$p^+ + p^+ \rightarrow 2D + e^+ + ve + 0.42$ MeV ... (1)

and

$p^+ + e^- + p^+ \rightarrow 2D + ve$ ... (2)

we get

$p^+ + e^- + \gamma \rightarrow n + ve$ ...... (3)

Reaction (1) is in my opinion completely unproven. There is an assumption that was made in the 1920s - a typical first-best-view - for lack of another. The corrected by means of a vague theory that a so-called tunnelling helped get state the desired reaction of deuterium plus fusion energy. The same can be said for reaction (2). In both cases also neutrinos, but where they came from - of that we do not know! In my reaction (3) fulfilled all the requirements for fundamental physics and logic. Baryon, lepton and charge number is correct. The energy produced when a neutron and a proton are united is the difference in mass times the speed of light squared (= 1.29 MeV).

What happens is thus that the electron (e^-) with its one component - a gamma photon - together with the gamma photon supplied in the reaction a W - boson (negatively charged intermediate vector boson) that quenches and neutralises the charge of the proton (p^+).

Thus, if (3) is a net reaction, so see the details in this way if we assume that step 1 is:

$e^- + \gamma = W^- + ve$ ... (4)

An incoming gamma photon (γ) takes care of the electron gamma photon, forming a negatively charged W^-. A neutrino (v_e) is released. Boson with charge -1 neutralise the proton charge +1 in (5).

$p^+ + W^- = n$ ... (5)

As a whole we get then (3). In some ways are this the *reverse* neutron decay! (The neutron decays into a boson and a proton, where the boson then shared in a gamma photon and an electron so that a neutrino formed.) The process exists evidently, is known since long and is called *electron capture (1)* which may thus be regarded as the *inverse* of neutron decay.

1. Electron capture is a process in which the proton- rich nucleus of an electrically neutral atom absorbs an inner atomic electron. (Wikipedia)

No proton is *transformed into a neutron.* A neutron is built up and is formed thanks to the boson that extinguish the electrostatic resistance of the proton, which takes up two opposite spinning photons - one from the electron and one from a gamma photon - and form a neutron. The Coulomb barrier ***ceases to exist***. This can be seen as the required catalytic effect.

In our case it holds that the electron indirectly play the role of catalyst and neutralises and extinguishes the protons positive charge once the negatively charged W·- boson become active and bound to the proton before it had time to decompose. No violence, but a process that occurs when all right conditions prevail. The photon — the gamma particle (γ) — combined with one of the electron's two components — a different picture than the common — which then connects to the proton to form a neutron. *A negative boson plus a proton thus provides a neutron.* At the same time released the neutrino which formed one of the two components which formed an electron (see graphic above)!

In the conventional approach, PEP-reaction of formula (2), thus requiring an extremely gravitational pressure and a temperature greater than 100 million degrees thus assumes that we in a first step get deuterium (2D). Deuterium is thus just like the proton a positive charge. The question of the origin of the neutrino will dim. It is in the detail that the truth is here. One cannot escape the feeling that they are simply not gotten together, but just invented! This is probably also the solution to the old riddle of the theory and observations do not match the number of observed neutrinos from the sun.

*

In my view of the Sun's gravity is that it *collects, arranges* and *organises* but **does *not*** do the ***job*** with the merger itself. The immensely power - ful forces and the high temperature of the sun's centre is not intended to exercise coercion and violence. The task is *not* to compress the hydrogen ions and overcome the Coulomb barrier, but to create the *conditions* necessary for a fusion reaction. The sun's centre is *not* a jumble of protons and electrons, as many believe, but they circulate very well organised in different paths with high speed and energy. Like we see on page 25. Sooner or later crossed their paths so that hydrogen and electrons plus photons can be merged into neutrons according to formula (3).

It is these courses with appropriate energies and intensities are supposed to consciously emulate here on Earth in special centres.

Schematic diagram of an electron (top) and a W-boson (bottom). Of the electron is assumed that the gray component is a photon, with spin 1. The other neutrino spin ½. Of the W boson is assumed to be two photons, but in the opposite spin (± 1), which also applies the electron.

It is this "coulomb barrier" in the formulas (1) and (2) that must be overcome in both the sun and in a merger process here on Earth is the basic idea since the 1920s. On the sun this is done, they say, by the large pressure and the extremely high temperatures. This calls for a temperature of 10^8 ° C, then 100 million degrees. Which no material can withstand without the need for a magnetic field that the process can be developed. And this has thus after 70 years as we have seen yet succeeded! The basic idea is that solar energy can only be tamed by a tremendous force! It is their view of nature: bam boom crash! It is applicable. So there is no scientific or experimental support to "(d) one-first step" to work. Seventy years of fruitless and a waste of time and resources in favour of this. That it would happen on the sun's just a guess and poor hypothesis.

In the current atomic model, the electron is basically a compact little ball whizzing around the nucleus. This mechanical "bullet" is only available as a statistical possibility of a "cloud".

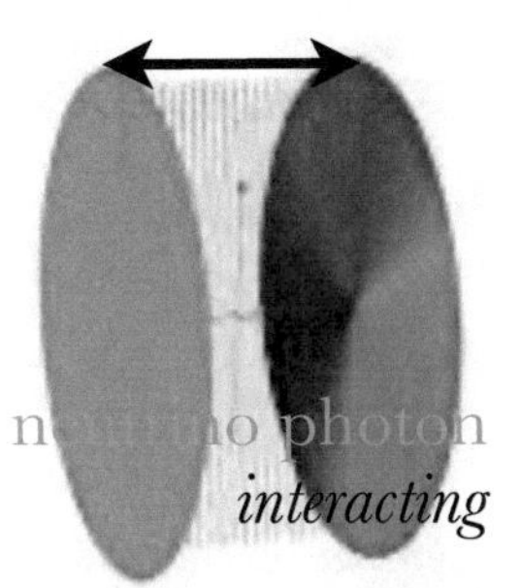

In this model (left), the electron thus has *two* components, one *neutrino* and one *photon.* They *interacts constantly.* In the sun comes the photon component connects to another incoming photon, an anti-photon (see reaction formula)! That new particle will connect to the proton whereby we get another new particle: a neutron. The neutrino (look at red circle above) is released.

Thus we see that in both the prevalent case and in this, form deuterium.In their case after the protons overcome the Coulomb barrier, is compressed to high pressure and temperature, and even perhaps reached a tunnel effect, if you will. "This first step is extremely slow," as they say. And good is it, otherwise the sun had burned up in a very short time. So then we have to ask ourselves what this means for our proposed process. Yes, it is sufficiently slow? What would make it slow? Let us consider the reaction (3) once:

$$p^+ + e^- + \gamma \text{ ---> } n + ve \text{ ..... } (3)$$

And note that from a conventional viewpoint it is hard to understand it. But we who have studied Part I of "Universum – utan Big Bang..." understand it very well. We know that a proton is shared set of three neutrinos, of which one is an antineutrino. One is a neutron (n), plus two coupled photons, a W - particle actually. A photon (γ) we see included in the formula, but where is the second photon? Well, it exists as one of the components of the electron. Will be removed, what remains? Well, a neutrino! (That's right, a neutrino and antineutrino not one.) The formula is correct. But why is the reaction so slow? Very conveniently, in terms of a photon of the exact size of the energy is present, plus this occurring when the two others also are. These meetings are rare. But still do it, otherwise the sun do not shine and burn today! Or what?

*

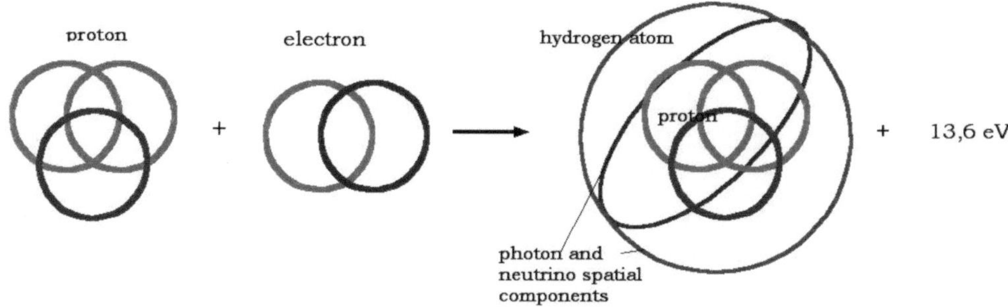

The new approach to the electron and its two components leads to a whole new atomic model.

Letter to a nuclear physicist. 2011

Hello,

Last summer I contacted you to discuss your amazing idea catalytic fusion. Not with muons, palladium, etc. but specifically with electrons. The same idea that I myself have used. You since 1996. You have taken the time to answer a couple of times:

Now lingers, however, you with an answer, Ok. But before you turn the page, so let me summarise the situation. But first ask: Why have you changed your mind regarding the catalytic fusion with electrons. To return to believing in tokamak technology? Very sad, I think. (Anyone who reads the criticism that is now star- ting to come to the ITER project and have enough reason to hesitate a little. See attached material).

Why you all in the whole world was contacted by me in this case thus depended on the catalytic fusion with electrons as catalysts, as I have already told. For me it was quite amazing that at least one person was on the same track. For, what I found was nobody else who dealt with this ver y issue, except perhaps an Ameri- can. I knew there were and are lots of people who would other wise treated right catalytic fusion (CF) with other catalysts than just electrons, so this idea is com- pletely unique. I will return to this.

However, I pointed out that our idea / hypothesis differ on closer inspection. We thus have different perceptions of the fundamental in the process. Actually not a bad thing but a good start at maybe something new and very good. None of us has claimed to be absolutely one hundred of his ground. We are both doubtful I would argue, perhaps even doubting. There's reason for it. Why would this parti- cular form of merger shall be that solves this apparently both incredibly difficult but also very urgent issue. As it invested so much resources on for so long? Why, for example, I imagine myself to have something to come by? An old retired far from all centres.

What I am saying, however, that (after detailed and lengthy studies on the subject): it my proposed device is worth trying and testing. For, it is based on an en- tirely new way to treat the problem. A new way of looking at the problem. A completely new concept. Which means that the probability of an energetically profitable outcome is very large.

I totally agree with you that the hypothesis I have so far presented can also be included in the ordinary quantum mechanics. So you have asked yourself: what new? And maybe prover pat a little. Well, what is new is that it pays off, despite what the mainstream reasoning about probabilities means. That's what I say. And it is not just about the method and the mechanisms underlying the process in question. The method and mechanisms are fundamental. But beyond that required something more.

All these fifty years of conscious brooding have brought me no nearer to the answer to the question, 'What are light quanta?' Nowadays every Tom, Dick and Harry thinks he knows it, but he is mistaken.
(Albert Einstein, in a (last?) letter to his old friend M A Besso, 1954)

I still do not believe that the statistical method of the Quantum Theory is the last word, but for the ti- me being I am alone in my opinion. (Albert Einstein, On Quantum Theory,1936)

Quantum theory is certainly imposing. But an inner voice tells me that it is not yet the real thing. Quantum theory says a lot, but does not really bring us any closer to the secret of the Old One. I, at any rate, am convinced that He (God) does not throw dice. (Albert Einstein, On Quantum Physics, Letter to Max Born, December 12, 1926)

The process can only be successful: Where and again: if you know what you are do- ing, because only then can the process be controlled and regulated sufficiently good way. This is thus fully possible today, unlike for 30-40-50 years ago. According to (our)? My recipe. Today it is sufficient computing power and the necessary technology at all. But not yesterday. I want to emphasise this. Yesterday was not the ne- cessary precision. And the opportunity. But now. It is important to detect and re- cognise this new. I know a thousand times all the usual objections to this: mathe- matical formalism of quantum mechanics does not allow a favourable outcome. The probability is too low. Etc, etc. Yes, I know this. But therein lies an errors in thinking. For, now is it about physics and not a game of chance and clumsy math!

I myself have had to deal with similar problems in the industry. It could take hours to process using a computer through self- regulation could control himself. Sometimes for days. But it came rather simple chemical processes. Physical of this kind requires much more computing power. On the other hand, we knew enough exactly what happened. We could fully trust await the maximum values (fine-tuning). We had a clear view of the process and therefore could control and regulate it successfully.

No mathematical formalism that obscured the reviews or gave us preconceptions.

This, then hook today, I mean. No one believes that it may be worthwhile because you only have blurry fuzzy images of what is actually happening. And so it is, and so it according to the standard theory in question. Probabilities for the entire unit and clouds of particles is what counts, not for individual particles and events. And the likelihood is saying that it does not pay.

It is very easy to misunderstand me here. I mean it is thus difficult enough to steer the process, SAS spot the right at the right time and so on. I speak thus also of probabilities, but of a different kind than the normal. But do you know what you are doing, so you can correctly set all the parameters and deploy all necessary resources. Resources that still are clean coffee money compared to what the tritium-powered tokamak mastodon in France requires. See attached article! Surely you must agree that 60 possibly 70 years of shortcomings must be due to something. Hey, hey, something must be wrong. All must admit this! There is something wrong! Could it be that there is no theory; the theory is incomplete or even wrong, or at least that the released energy becomes so expensive that an economic exploitation of controlled thermonuclear reactions are not possible in a reasonable time. Who says that? Well, at least two Nobel Prize winners Albert Einstein and Hannes Alfvén. But also "hydrogen bomb father" who for many years worked with projects to try to tame this bomb: Edward Teller. And many ma- ny more, of course.

Regarding this question, I say that the prevalent theories unfortunately is not only extremely weak; the worst part is that it is directly misleading. It has misled researchers in 60-70-80 years. It gives all the wrong models and images of what is actually happening. This is evident, for example, the idea of using muons. (Muons considered to be the lie closer to the nucleus. And therefore better dampen the Coulomb effect for a longer time than the much lighter electrons. Or something like that idea seems to be). Or the belief in tokamak technology where it is considered necessary to first heat up a plasma to several million degrees! And thus believe they mimic what happens in the sun. Or head the idea that the maximum force (millions of degrees, enormously powerful electric discharges and huge chaotic magnetic fields, etc.) are needed to "tame" nature and its forces. I could say a lot about this, but now I want to round off.

What researchers are advocating a further round of merger attempt with the same old technology with the device ITER in France? That ignorant politicians have be- en lulled into the belief that this road is the only one, is not so strange. Energy is of course a key issue for any society. But all naturalists should stand up and condemn this waste (cheating)? Of public funds. One must realise that ended in a final stage of gigantism – of the cathedral building. And that a new theory, as the late Alfvén said long ago. I also say that a good friend of mine said, it is reminiscent of the situation in which the child in the story said: Emperor is naked!

*

My recipe is thus that. Make sure to hold two opposing streams of deuterons (D +) which is circulated at an appropriate speed, frequency and intensity. Let them come on a collision course. Sort then by one or two streams of electrons at a sui- table rate, frequency and intensity which may neutralise or excite specific ions in the stream just before collision. It is of course also to ensure that an ion is ne- gatively charged another positively. A cation must meet an anion. The various states of each particle may be a number of nanoseconds or femtoseconds. Instruments must measure the outcome. The possibilities are endless and many variations, increasing the chances the successful merger. Then allow computers to control and regulate these flows, parameters and variables to a maximum outcome. When all cross sections are well known is the probability of successful reunion easy to calculate. Make sure it all gets a ceramic enclosure that can withstand up to thousand degrees. Let the generated heat to warm the water to drive steam turbines, etc. (As you probably understand, this technology has nothing to do with so-called "cold fusion!" Absolutely not.)

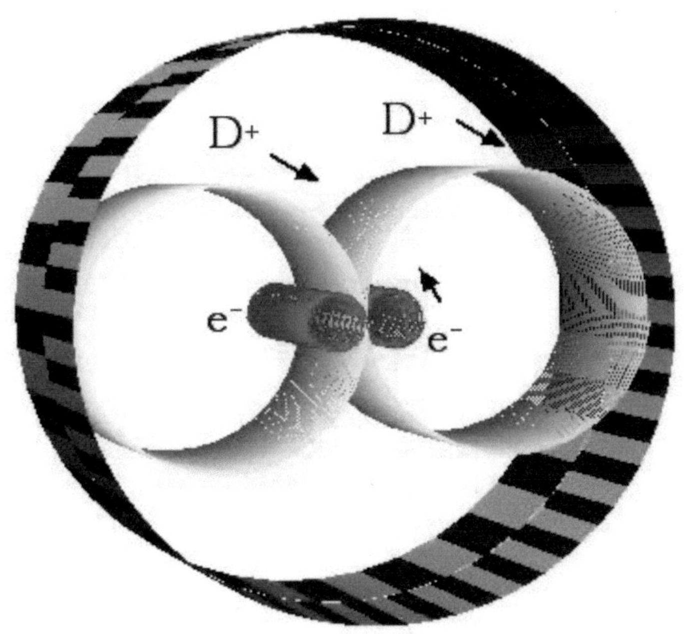

$D^+ + D^+ +$ >>> He2+ + 2e + gamma ray

Coulomb forces is apparently no obstacles or problems with this device, on the contrary, their pull effect is utilised for control. Since all parameters are known and can be adjusted, next to a degree of uncertainty, the process can also be simulated using computer program. (The Tokamak's chaotic behaviour is an obstacle to such.) Here we see besides that the process works catalytically, then basically the same electrons can be used again and again...

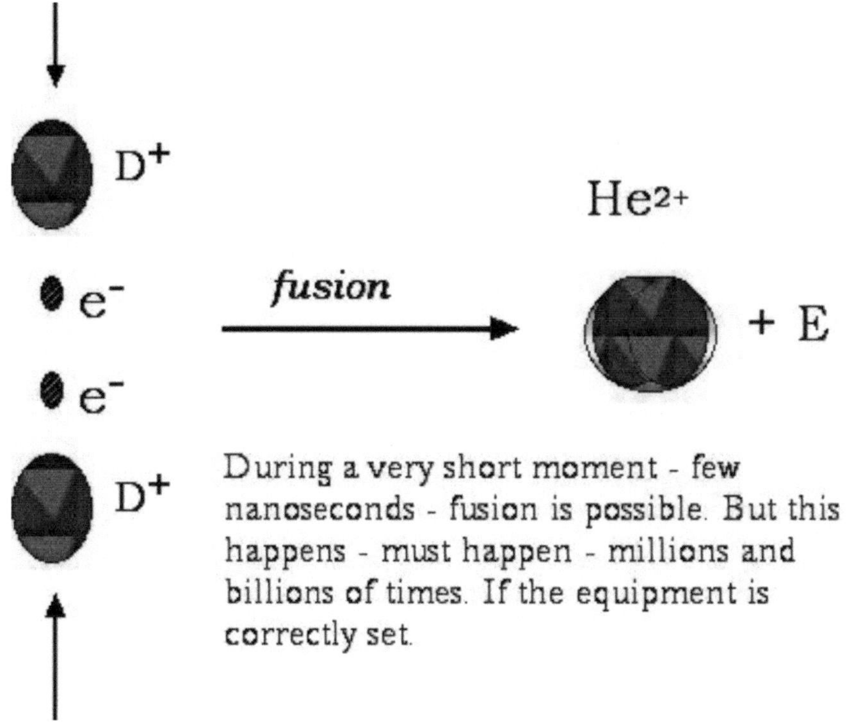

The cost of a device becomes as said minimal compared to the ITER project, especially for a test and experimental apparatus. Moreover pulled the emergency brake now in the runaway costs of the ITER project. The risk is high that the project is not completed.

Seems my proposal to the solution of the fusion problem easily; too simple? Do not worry, it will be very difficult to control and regulate the process with the necessary precision. But when you do not grope blindly without knowing what you are doing and the whole thing is to some extent self-regulating and self-governing, one can confidently work on and wait for the desired level of outcome. And by all means help of course. Much to analyse and discuss, change and improve so on. Something you with some patience can eventually get control of. Perhaps with the help of simulation programs. Road to controlled fusion at a reasonable cost within a relatively short time is thus possible. Playing it matter to you then may amend and revise some preconceived ideas about how it could possibly work? When reality intrudes on. Experiments and observations must surely get to decide what is right or wrong.

Or perhaps there is something fundamentally wrong and someone really serious hitch in my proposal and outline? Everything is obviously not right and proper, but I wonder if there is something truly significant error. Would be grateful if someone in that case can enlighten me.

Finally. The solution to the fusion problem is not finding a shrewd, slick and clever new mechanism that nobody else previously come on. I think not, rather it is important to see what is already available and put it in a fruitful context. And develop it into an art. Why it anyway constantly pops up such attempts to find a clever mechanism, depends on – if someone asks me – that one way or another misunderstanding is what this is about.

Sincerely,
Åke Hedberg

Hello again,

You hesitate, I guess. You might not care to answer this time. I understand; I might have offended you in any way; You have the right to doubt what I say about electrons and atoms, fusion technology, and more.
But I cannot help that I discovered that the electron is a neutrino bound to a photon. Hence its properties. Nor yards I to proton is three neutrinos linked. Or that the neutron is two photons bound to three neutrinos (= proton). Hence the proton, respectively. The free neutron properties. (Anti-forms means that neutrinos here is exchanged for anti-neutrinos.)
In summary:

The electron is a neutrino bonded to a photon. A free neutrino as a ditto photos have no electric charge. In the bound state, however, the neutrino and photon as one part have charging.

The proton is three neutrinos linked. They got thirds charges.

The neutron are two photons (in the opposite spin = -1) bonded to three neutrinos.

The photons neutralises neutrinos / proton charges (+1).

Atom thus have a different structure than the usual "planetary" and "Schrödinger," , etc., the hydrogen atom nucleus will thus contain, consist of four neutrinos. "Electron" -the neutrino, which is also bound to a photon neutralise the proton's charge.

Based on the known data on the electron, proton and neutron resting masses, frequencies, and Planck's constant h, it is easy to calculate how "hard" these bonds are. And much more interesting. Not least is the structure of these particles to visualise, thus made intelligible in a completely different way than usual.

Study e.g. the free neutron decay in two steps. In a first step of a proton and a boson. Thus two products.

Step 1:
n >>> p + + W- (= 2 photons).

The W- bosons secret is that there are two photons bound to each other with opposite spin. The charge is -1. The constellation is unstable. It decays rapidly into an electron and an anti-neutrino. Thus two products again.

Step 2:
W >>> e + ν (= anti-neutrino)

Policy:
One (1) photon can be "split" in a neutrino and an anti-neutrino. This is thus in this case, one of which binds to a neutrino images, and becomes an electron. The second — the antineutrino thus — will be free. W boson disappears.

neutrino + antineutrino >>> may through annihilation formed into a photon.

a photon can be in the opposite process through defined pair formation as a neutrino + an antineutrino.

Neutrinos and photons can in different ways under certain circumstances and in different proportions shaping electrons, protons, neutrons and other particles.

I cannot see anything that is contrary to fundamental physics.

Finally.

"(A) support for costly experiments of your kind, I cannot give," writes you. I have not asked. For me this is mostly a fun thing, it's very interesting to me that after many decades of studies finally begin to understand how things fit in the crowd. It has for me been very satisfactory. Really regret those still treading out the grain of the usual swamp, maze or thicket.

But speaking of this, you need enough not to worry the worst. For, first, what I can see, you need a test facility only cost a fraction of a fraction of what Tokamak experiments costs each day today. (Experiments that are guaranteed not going to get any pleasure from other than those daily can make money on the project, of course. It purely scientific value is additionally entirely negative. It felt your colleague H. Alfvén already for more than 25 years ago. And he warned such as Chernobyl and Japan. Right?).

After what has now happened in Japan, does not take much more can be said than that the energy crisis is acute. But I need an authority that can back up, one that is at least not entirely negative.

My greetings
Åke Hedberg

*

In summary and conclusion
Our nuclear scientists, then, with that my view of Coulomb forces can "be included in the ordinary quantum mechanics" but only after a sliding scale. My belief is that these forces and the impact they make is to be regarded as a quantum phenomenon — the Coulomb forces is on or off. Zero or one, plus or minus.

And it is during this brief moment they are, they are zero, the merger can take place. How short it is, then it is only a technical issue and a matter of precision that use of the time available. The fusion process is, as I said, to some extent, thus both autonomous and self-regulating! This is thanks to the attractive coulomb forces! A technique with feeling and precision, however, must be developed into a kind of art.

Please note that: "The process can only be successful: If and again: if you know what you are doing (which by definition, not a" statistical "quantum mechanics does) because then only the process can be controlled and regulated sufficiently good way. This is thus fully possible today, unlike for 30-40-50 years ago. According to (our)? My recipe. *Today it is sufficient computing power and the necessary technology.* But not yesterday. I would again emphasise this. Yesterday was not the necessary precision. And the opportunity. But now. It is important to detect and recognise this new. ("Quote from letter to a nuclear physicist", January 2011.)

If this new theory applies, then the solution rather trivial. But since the nuclear scientists we are dealing with apparently no discovery, realised or understood "this new" and this potential art and since he has not been in touch for so long so I go now on. Instead, he perseveres, departing from its previous ideas and theories of catalysis by the electrons to once again support the decay to the old Tokamak technology.

Key words are thus *knowledge*, *technology* and *art*. After decades of failure despite the billions of dollars, rubels and euros, huge almost unlimited technical resources and thousands and thousands of scientists so it should be obvious that they lack, it is requisite knowledge.

Therefore they are not capable of applying and developing the technology needed to make it work; and least able to develop this into an art. And it is clear that as long as our nuclear physicist and his colleagues with his barren, statistical and mechanical approach may continue to dominate the natural sciences will no solution of the fusion problem to occur.

On the contrary, humanity eternal times get stuck with coal and oil, and not least the dangerous and extremely costly nuclear power and thus exacerbate environmental and energy crisis, something that the current world order, not for long will hold together. Not even the 30-50 year they set a limit for a possible solution.

***

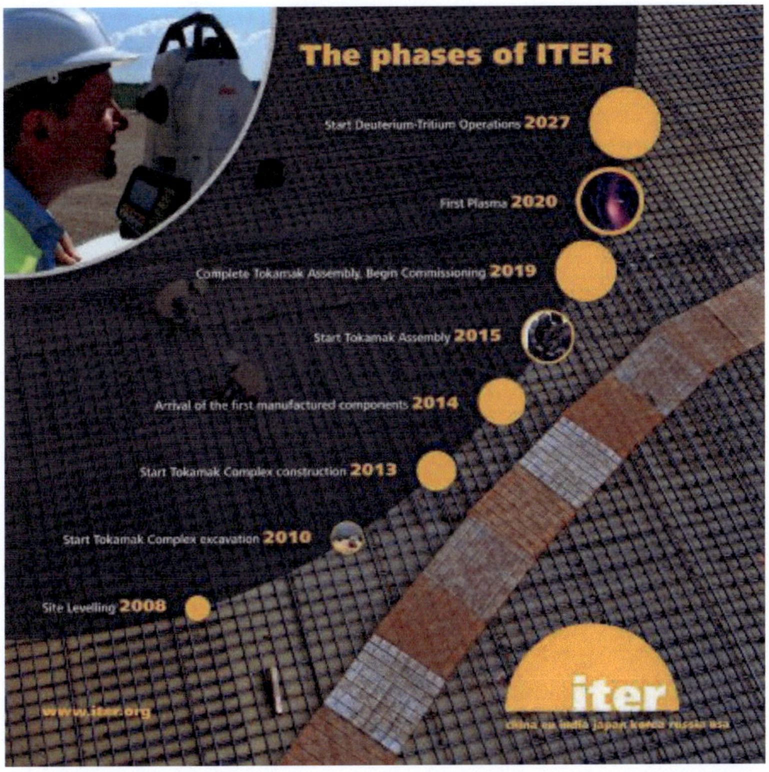

Common theory about what happens on the Sun:

(Which is based on the fact that it is the extremely high temperature and ditto pressure that does the trick).

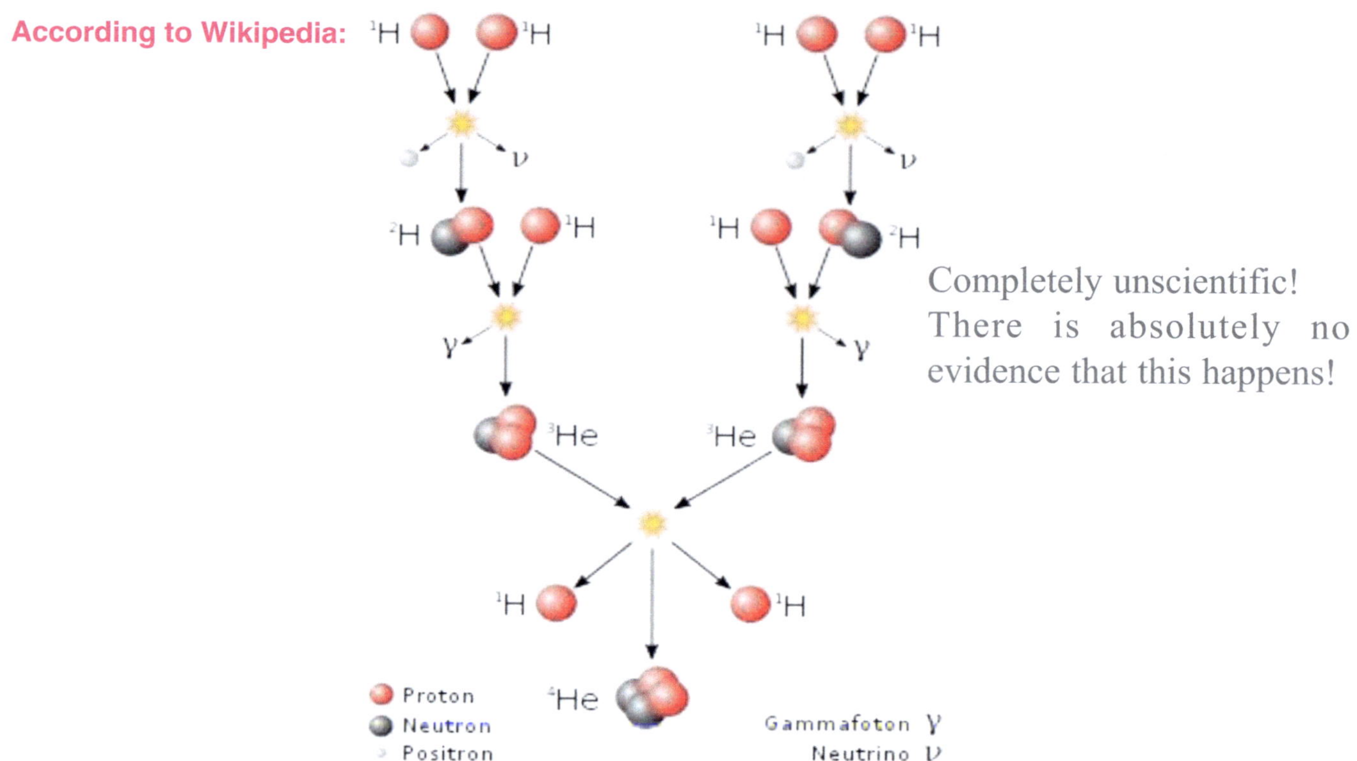

The first step involves the fusion of two hydrogen nuclei 1H (protons) to deuterium, which releases a positron and a neutrino because a proton becomes a neutron.

$$p^+ + p^+ \rightarrow {}^2D + e^+ + \nu_e + 0{,}42\ \text{MeV}$$

This first step is extremely slow, both because the protons have to tunnel through the Coulomb barrier and because it depends on the weak interaction.

The positron is annihilated immediately with an electron and their mass energy is carried away by two gamma photons.

$$e^- + e^+ \rightarrow 2\,\gamma + 1{,}02\ \text{MeV}$$

After this, the deuterium created in the first step can fuse with another hydrogen nucleus to form a light isotope of helium, 3 He.

$$^2D + {}^1H \rightarrow {}^3He + \gamma + 5{,}49\ \text{MeV}$$

Instead, neutrons n, are formed in a first step.
These neutrons then form deuterium, 2D+ Neutrons ·
are formed when its decay instead means a build-up of neutrons. (Wikipedia)

The new theory of what happens on the Sun.
(Which is based on a more fruitful and sophisticated theory).

Att that neutrons would be formed in the "fusion of two hydrogen nuclei", is just one completely unproven statement. Let alone that positrons and neutrinos are formed in such an event. No have ever been able to observe anywhere via experiments or other types of events such reactions. One wonders why they have to invent such types of reactions. (See my explanation on next page!). What happens instead in the reactions reported in the following, is documented and proven in countless experiments and observations.
But first some background. Let's study what happens in an ordinary decay of neutrons:

n — > p+ +W- (vector boson)

W- — > e- + vanti (antineutrino)

It is a well known fact that the reaction is reversible. In other words:

p+ + e- + vanti — > n

Or

p+ + W- — > n

Of course, such a reaction is quite rare, it requires e.g. that the neutrino has the right frequency and is in the right place at the right time. But it is as it should be, otherwise the following reaction would have been too frequent. And the sun would have exploded. Thanks to the fact that the reaction is thus rare, the sun can shine and warm at a leisurely pace.

n + p+ — > D+

Etc. Now, in a similar way, deuterium and tritium can constantly be formed in the Sun and the other stars and thus give rise to all the necessary fusion reactions. The reactions can take place at the right pace so that the sun can shine and warm us. If you want to guess, the reaction is probably catalyzed by electrons, of which there are plenty throughout the solar sphere. As we can see, no extra neutrinos are formed in the process, which it does in the usual one. On the contrary, the process requires an addition of neutrinos. Where do they come from? From outer space, of course. This theory formation thus provides a clear answer and logical solution to *the neutrino problem!* (In the usual theory, the "solution" is to blame it on the neutrinos themselves who are the problem, they change character like Superman or similar on their journey from the Sun, and therefore the measurements are misinterpreted, they claim).

Neutrinos and anti-neutrinos can thus both be created and destroyed, but preferably together with their charged partner the antielectron/electron. That neutrinos can affect and combine with other particles is well known. The reaction below is used, for example, at the Sypole to detect neutrinos.

v + n — > p+ + e-

See: (https://en.wikipedia.org/wiki/Sudbury_Neutrino_Observatory#Charged_current_interaction).

***

The principle of the Fusion Process, catalyzed and computer controlled.

The two proton cannons shoot deuterium ions (D+) at each other. The two electron cannons do the same, but with ordinary electrons. But not anyway. The ideal situation occurs when, thanks to the electrons, one of the D+ gets a negative charge (D-), while the other keeps its positive, whereby the two will attract each other and unite to form a helium – He. This merging – fusion – of the two oppositely charged deuterium ions into helium produces a large excess of energy in the form of gamma radiation. The electrons thus play the role of catalysts. If the particles miss each other, they are repelled and return. Repeatedly. The process is controlled and regulated with computers.

$$D^+ + D^+ + 2e^- \longrightarrow 2H + 2H = He^{2+} + \text{energy}$$

This experimental facility can be varied in many different ways. With different angles, different number of cannons and charges, different energy levels, different frequencies, etc.

$$\text{Or} = 2D^+ + 2e^- = 2\ 2H + \text{energy}$$

The processes should take place under as great a vacuum as possible.

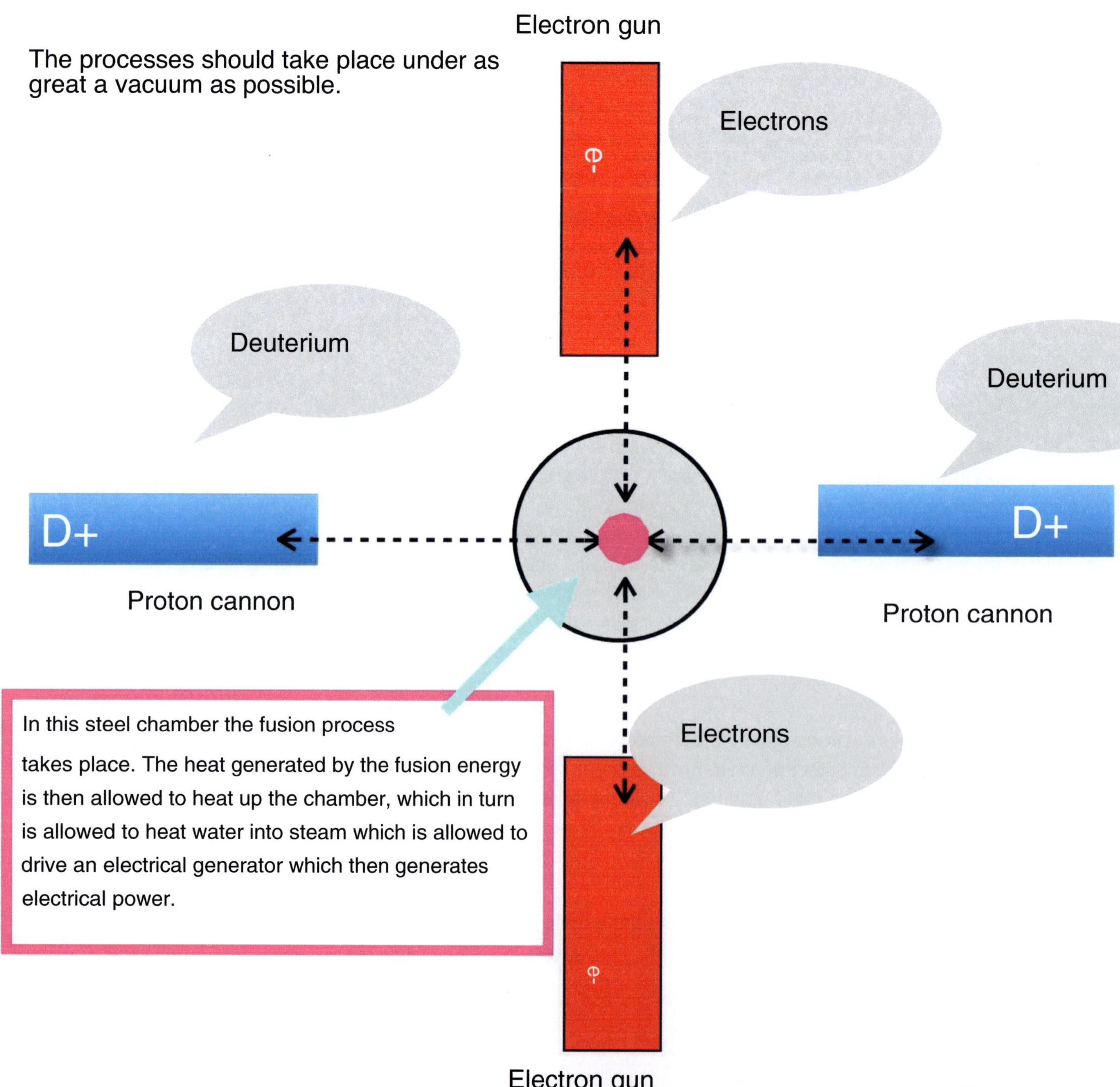

Fusion power

The Joint European Torus (JET) magnetic fusion experiment in 1991

Fusion power is a proposed form of power generation that would generate electricity by using heat from nuclear fusion reactions. In a fusion process, two lighter atomic nuclei combine to form a heavier nucleus, while releasing energy. Devices designed to harness this energy are known as fusion reactors. Research into fusion reactors began in the 1940s, but as of 2024, no device has reached net power, although net positive reactions have been achieved.[1][2][3][4]

Fusion processes require fuel and a confined environment with sufficient temperature, pressure, and confinement time to create a plasma in which fusion can occur. The combination of these figures that results in a power-producing system is known as the Lawson criterion. In stars the most common fuel is hydrogen, and gravity provides extremely long confinement times that reach the conditions needed for fusion energy production. Proposed fusion reactors generally use heavy hydrogen isotopes such as deuterium and tritium (and especially a mixture of the two), which react more easily than protium (the most common hydrogen isotope) and produce a helium nucleus and an energized neutron,[5] to allow them to reach the Lawson criterion requirements with less extreme conditions. Most designs aim to heat their fuel to around 100 million kelvins, which presents a major challenge in producing a successful design. Tritium is extremely rare on earth, having a half life of only ~12.3 years. Consequently, during the operation of envisioned fusion reactors, known as breeder reactors, helium cooled pebble beds (HCPBs) are subjected to neutron fluxes to generate tritium to complete the fuel cycle.[6]

As a source of power, nuclear fusion has a number of potential advantages compared to fission. These include reduced radioactivity in operation, little high-level nuclear waste, ample fuel supplies (assuming tritium breeding or some forms of aneutronic fuels), and increased safety. However, the necessary combination of temperature, pressure, and duration has proven to be difficult to produce in a practical and economical manner. A second issue that affects common reactions is managing neutrons that are released during the reaction, which over time degrade many common materials used within the reaction chamber.

Fusion researchers have investigated various confinement concepts. The early emphasis was on three main systems: z-pinch, stellarator, and magnetic mirror. The current leading designs are the tokamak and inertial confinement (ICF) by laser. Both designs are under research at very large scales, most notably the ITER tokamak in France and the National Ignition Facility (NIF) laser in the United States. Researchers are also studying other designs that may offer less expensive approaches. Among these alternatives, there is increasing interest in magnetized target fusion and inertial electrostatic confinement, and new variations of the stellarator.

Background

Mechanism

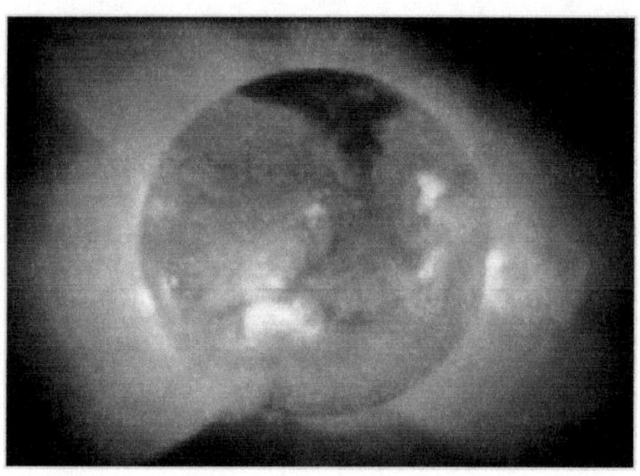

The Sun, like other stars, is a natural fusion reactor, where stellar nucleosynthesis transforms lighter elements into heavier elements with the release of energy.

Fusion reactions occur when two or more atomic nuclei come close enough for long enough that the nuclear force pulling them together exceeds the electrostatic force pushing them apart, fusing them into heavier nuclei. For nuclei heavier than iron-56, the reaction is endothermic, requiring an input of energy.[7] The heavy nuclei bigger than iron have many more protons resulting in a greater repulsive force. For nuclei lighter than iron-56, the reaction is exothermic, releasing energy when they fuse. Since hydrogen has a single proton in its nucleus, it requires the least effort to attain fusion, and yields the most net energy output. Also since it has one electron, hydrogen is the easiest fuel to fully ionize.

The repulsive electrostatic interaction between nuclei operates across larger distances than the strong force, which has a range of roughly one femtometer—the diameter of a proton or neutron. The fuel atoms must be supplied enough kinetic energy to approach one another closely enough for the strong force to overcome the electrostatic repulsion in order to initiate fusion. The "Coulomb barrier" is the quantity of kinetic energy required to move the fuel atoms near enough. Atoms can be heated to extremely high temperatures or accelerated in a particle accelerator to produce this energy.

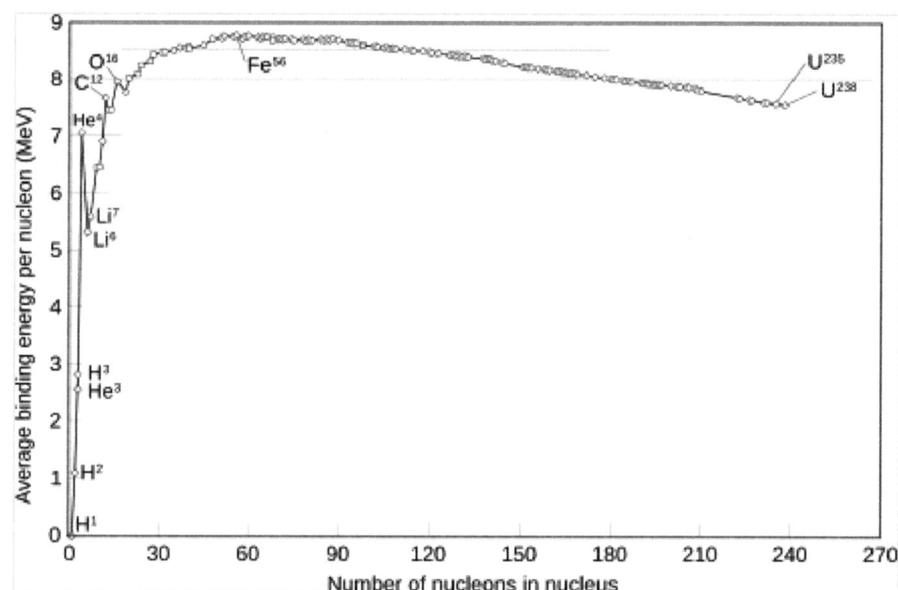

Binding energy for different atomic nuclei. Iron-56 has the highest, making it the most stable. Nuclei to the left are likely to release energy when they fuse (fusion); those to the far right are likely to be unstable and release energy when they split (fission).

An atom loses its electrons once it is heated past its ionization energy. An ion is the name for the resultant bare nucleus. The result of this ionization is plasma, which is a heated cloud of ions and free electrons that were formerly bound to them. Plasmas are electrically conducting and magnetically controlled because the charges are separated. This is used by several fusion devices to confine the hot particles.

Contact

akehedberg2015@gmail.com

***

Why have thousands of scientists and technicians with access to huge material and financial resources and over 60 – 70 years of extremely costly experiments, not for practical pur poses been able to solve the problem of so-called controlled fusion? And imitate the natural processes occurring on the sun? Thus a " calm" fusion of hydrogen isotopes into helium, etc. Why do they have so thoroughly failed?

Yes, why have they failed to resolve the issue of energy and thus rid the world of all catastrophic problems with nuclear power, oil and coal?

These are some of the most important issues the author of this essay sets.

But he also provides answers to both theoretical and technical problems and not least the fundamental solutions. And why the theory of what happens on the sun does not match the actual observations.

Well, the future ITER project in France? Will it succeed? There is not the slightest chance! Precisely because it is based on exactly the same 70-year-old principles ...the principles of the TOKAMAK.

The deeper and most obvious reason for the long-standing, incorrect theories that exist about what is happening on our sun and the other stars is that the basic theory about it is ***not included*** in the ***training*** of nuclear and plasma physicists. They therefore do not have the required skills. And that they tragically do not understand this.

So it is not primarily the politicians who are responsible for the climate crisis (as Greta believes), but the responsibility lies with these (unlearned) physicists

***